GROWING MARIJUANA OUTDOORS

The Ultimate Guide for Beginners

By

Micheal L.Smith

Copyright © 2022 Micheal L.Smith

All rights reserved. No part of this book may be reproduced or transmitted in any form or by any means, electronic or mechanical, including photocopying, recording, or by any information storage and retrieval system, without permission in writing from the author! This book is a work of non-fiction. The views
expressed are solely those of the author and do not necessarily reflect the views of the publisher, and the publisher hereby disclaims any responsibility for them

CONTENTS

CHAPTER 1	**4**
Introduction to Growing Marijuana Outdoors	4
CHAPTER 2	**8**
Choosing the Right Location for Growing Marijuana Outdoors	8
CHAPTER 3	**11**
Understanding the Different Types of Outdoor Cannabis Strains	11
CHAPTER 4	**17**
Selecting the Right Soil for Outdoor Cannabis Growth	17
CHAPTER 5	**25**
Nutrient Requirements for Outdoor Cannabis Growth	25
CHAPTER 6	**27**
Watering and Feeding Your Outdoor Cannabis Plants	27
CHAPTER 7	**32**
Understanding Light and Shade Requirements for Outdoor Cannabis Growth	32
CHAPTER 8	**37**
Dealing with Outdoor Cannabis Pests and Diseases	37
CHAPTER 9	**40**
Plant Training Techniques for Maximum Outdoor Cannabis Yields	40
CHAPTER 10	**45**
Harvesting, Drying, and Curing Outdoor Cannabis	45

CHAPTER 11   **50**
Common Outdoor Cannabis Growing Mistakes to Avoid   50

CHAPTER 12   **53**
Troubleshooting Common Outdoor Cannabis Growing Problems   53

CHAPTER 13   **56**
Maximizing Your Outdoor Cannabis Growing Experience   56

CHAPTER 14   **58**
Conclusion   58

CHAPTER 1

Introduction to Growing Marijuana Outdoors

Growing marijuana outdoors is an exciting and rewarding experience. It can be a great way to get yourself closer to nature, while also enjoying the benefits of growing your cannabis. Growing marijuana outdoors has several advantages that can be beneficial to both the grower and the plants.

First, growing outdoors is much more cost-effective than indoor gardening. Outdoor marijuana plants do not require the added expense of grow lights, ventilation systems, and other indoor gardening equipment which can be costly.

Furthermore, since outdoor plants are exposed to natural sunlight and rain, there is no need for additional irrigation costs or energy bills associated with artificially providing the plant's light and water needs.

Second, outdoor marijuana plants tend to produce higher yields than indoor plants. This is because outdoor plants have larger root systems, which allow them to access more nutrients and water from the soil, as well as larger flower heads due to increased exposure to sunlight. The increased flowering also allows for higher THC levels, as well as a greater variety of strains.

Third, outdoor marijuana plants tend to be stronger and more resilient than indoor plants. This is because they are exposed to harsher weather conditions, such as wind, rain, and UV rays. As a result, they tend to develop stronger stems, thicker leaves, and larger roots which make them more resistant to disease and pests.

Fourth, outdoor marijuana plants are less prone to molds and mildew. This is because outdoor plants are exposed to more air movement, which helps to keep the humidity levels lower and reduce the risk of molds and mildew.

Finally, outdoor marijuana plants are less likely to be affected by fluctuations in temperature and light. Since the plants are exposed to natural sunlight and rain, they can better adjust to changing environmental conditions, making them less likely to suffer from stunted growth or other issues related to temperature and light fluctuations.

All in all, growing marijuana outdoors has many advantages for both the grower and the plants. Not only is it cost-effective, but it also provides higher yields, greater resiliency, and fewer risks of molds and mildew.

Furthermore, outdoor plants can better adjust to changing environmental conditions, making them less likely to suffer from stunted growth or other issues. With all these benefits, it's easy to see why many growers prefer to grow outdoors.

One of the main benefits of growing marijuana outdoors is the cost savings. You don't need to invest in expensive grow lights or other indoor growing equipment, and you don't have to worry about power bills either. You can also take advantage of the natural environment, such as natural sunlight and rainfall, to help your plants thrive.

When it comes to growing marijuana outdoors, it's important to keep in mind the fact that it's a living organism and requires care to thrive. You need to consider the climate, humidity, sunlight, soil type, and other factors that can affect the growth of your plants.It's also important to choose the right strain for your outdoor environment.

This will ensure that your plants receive the right nutrients and conditions to thrive.When you're ready to start growing marijuana outdoors, there are a few steps you can take to ensure success.

First, you should choose an area that is sheltered from wind, rain, and other elements. You also want to make sure that you're planting in an area with plenty of direct sunlight. Next, you'll need to select the right soil for your plants and properly prepare the soil with nutrients and amendments.

Finally, you'll need to monitor the plants throughout the growing process, making sure that they receive the right amount of sunlight, water, and nutrients. With the right approach, you'll be able to produce a healthy, high-quality harvest of marijuana that you can enjoy for years to come.

As we go on you will get to know the best tips and tricks for growing marijuana outdoors, from planting to harvesting, so you can take advantage of the natural environment and enjoy the fruits of your labor.

CHAPTER 2

Choosing the Right Location for Growing Marijuana Outdoors

When it comes to growing marijuana outdoors, selecting the appropriate location is essential for successful growth. There are numerous factors to consider such as:

1. Sunlight: Marijuana plants need a minimum of 8 hours of direct sunlight each day to grow and thrive. If your location receives too much shade during the day, it will limit the quality and quantity of your harvest.

Additionally, the sun's intensity can vary greatly by location. A location that receives too much intense sunlight can cause the plants to wilt and dry out, while a location that receives too little sunlight can cause the plants to stretch and become weak.

2. Temperature: The ideal temperature for growing marijuana outdoors is between 65 to 80 degrees Fahrenheit. If the temperature falls below this range, the plants may become stunted and unhealthy. Too much heat can also be detrimental to the plants as it can cause them to dry out and become less productive.

3. Wind: Wind can be beneficial for outdoor marijuana grows as it helps to keep the plants cool and can prevent mold from forming on the leaves. However, too much wind can cause the plants to become stressed and can damage the delicate leaves and buds. Therefore, a location that is sheltered from strong winds is ideal.

4. Soil: Marijuana plants need well-draining, nutrient-rich soil to thrive. The soil should be light and fluffy with a slightly acidic pH level. Poorly drained soil can cause the plants to become waterlogged and can lead to root rot.

5. Water: Water is essential for any marijuana growth, and the plants will require regular watering. Look for a location close to a water source, such as a river, lake, or stream, to ensure that the plants have easy access to water. Additionally, the soil should not be too wet or too dry; it should be moist but not soggy.

6. Local laws and regulations: Before beginning a cannabis cultivation operation, it is important to research and understand the local laws and regulations for growing marijuana. Different municipalities have different rules and regulations regarding outdoor cultivation, such as setbacks from neighboring properties, fencing requirements, and zoning restrictions.

7. Security: Outdoor cultivation requires a secure and private location to protect plants from theft and contamination.

8. Pest control: Depending on the location, there may be an increased risk of pests or diseases that could damage the crop. Proper pest control measures should be taken to ensure a successful harvest.

9. Cost: Costs associated with the site, such as rent, utilities, and labor, should be taken into account when selecting a location for outdoor cultivation.

10. Accessibility: Outdoor cultivation sites should be easily accessible for workers and delivery vehicles.

By taking all of these factors into consideration, you can choose the perfect location for your outdoor marijuana to grow. With the right location, you will be well on your way to a successful and abundant harvest.

CHAPTER 3

Understanding the Different Types of Outdoor Cannabis Strains

When it comes to outdoor cannabis strains, there are many different types to choose from. Each type of strain has its unique characteristics and benefits, so understanding the differences between them is essential for getting the most out of your outdoor growing experience.

First, let's start with Sativa a type of cannabis plant that is commonly grown outdoors. It is known for its tall, slender, and light green leaves, and its long flowering period.

Sativa plants are usually taller and can grow to be up to 20 feet in height, while Indica plants tend to be shorter, and typically only grow to be about 6 feet in height.

Sativa plants are popular for their high-energy effects and are often used by those looking for an uplifting, energizing, and creative high. They tend to have a more energizing and uplifting effect than Indica strains, which have a more sedative and relaxing effect.

Sativa plants are typically grown in warmer climates and tend to flower for longer periods than other cannabis plants.

They can take up to 10-12 weeks to finish flowering and need more sunlight than other cannabis plants. Sativa plants are also more resistant to mold and pests than other cannabis plants, making them a great choice for outdoor growers.

Sativa plants tend to produce more THC than other cannabis plants. This can lead to a more intense and potent high. As such, they are often used by experienced cannabis users who are looking for a more intense experience.

Sativa plants are popular among outdoor growers because of their long flowering period and high THC levels. They can also be grown indoors, but require more space than other cannabis plants.

Overall, Sativa is a great choice for outdoor growers who are looking for an energizing and uplifting high. They require more time and effort to grow, but the reward is well worth it!

Next, let's look at Indica. Indica is a variety of cannabis that has been used for centuries in Asia, Central Asia, and the Middle East. It is a short, bushy plant that typically grows to a height of two to six feet and has broad, dark green leaves. Its buds are usually dense and covered with a thick layer of resin, giving them a sticky texture.

Indica is known for its tranquilizing and sedative effects, which make it popular for medical and recreational use. Its effects are typically felt more in the body rather than in the mind, making it ideal for treating pain, insomnia, and anxiety.

When grown outdoors, Indica can thrive in a variety of climates, from hot and dry to cooler and more humid. It is also relatively easy to grow, needing only moderate amounts of sunlight, water, and nutrients.

Indica plants tend to be more resistant to pests and diseases than other cannabis varieties, making them easier to cultivate. They also tend to be shorter than other varieties, which can be beneficial for those living in areas with strict height restrictions.

Indica can also be used to make a variety of cannabis products, such as oils, edibles, and concentrates. Its high THC content makes it ideal for making potent extracts and tinctures, which can be used for both medical and recreational purposes.

In conclusion, Indica is a hardy and versatile strain of cannabis that can be used for a wide range of applications. Its sedative effects make it ideal for treating pain and anxiety, while its shorter height and resistance to pests and diseases make it easy to cultivate for both indoor and outdoor growing. Its high THC content also makes it ideal for making potent extracts and tinctures.

Finally, let's look at Hybrid strains. Hybrid cannabis strains are a type of cannabis strain that is a combination of two or more different cannabis strains. Hybrid cannabis strains are created to bring together the best characteristics of each parent strain, such as flavor, aroma, potency, and effects.

Hybrid strains can be either Indica or Sativa-dominant, as well as a combination of both, yielding an effect that is neither purely Indica nor Sativa.

Outdoor cannabis is cannabis that is grown outdoors instead of in a controlled indoor environment. Outdoor cannabis is typically grown in climates that are conducive to the growth of cannabis, such as the Mediterranean, Central and South America, and parts of the United States.

Outdoor cannabis grows differently than indoor cannabis, often yielding larger plants and more abundant harvests. Outdoor cannabis is also known to have a more natural flavor and aroma than indoor cannabis.

Hybrid outdoor cannabis strains are a combination of two or more outdoor cannabis strains that have been bred together to bring out the best qualities of each parent strain.

The effects of these hybrid outdoor cannabis strains can range from relaxing, euphoric, and sedative to energizing, uplifting, and creative, depending on the parent strains.

The flavors and aromas of these hybrid outdoor cannabis strains can range from sweet and fruity to earthy and musky, again depending on the parent strains.

The popularity of hybrid outdoor cannabis strains is increasing due to the variety of effects that they offer. Outdoor cannabis strains are ideal for outdoor growers looking to maximize their yield and complexity of flavor and aroma, as well as for medical cannabis users and recreational users alike.

Hybrid outdoor cannabis strains offer growers the opportunity to customize their growing conditions to achieve the desired flavors and effects.

Whether you are a medical cannabis user or a recreational cannabis user, hybrid outdoor cannabis strains offer a wealth of choices. With a wide range of effects, flavors, and aromas, hybrid outdoor cannabis strains offer something for everyone.

Now that you understand the different types of outdoor cannabis strains, you can start to make an informed decision about which strain is right for you.

Each strain has its unique benefits and characteristics, so it's important to do your research before planting.

By understanding the differences between the various types of outdoor cannabis strains, you can ensure that you get the most out of your outdoor growing experience.

CHAPTER 4

Selecting the Right Soil for Outdoor Cannabis Growth

When it comes to outdoor cannabis growth, selecting the right soil is one of the most important decisions you can make. The soil you choose will determine how your plants will grow and how much yield you can get from them. Proper soil selection will also help you avoid common problems such as nutrient deficiencies and pest infestations.

When choosing soil for outdoor cannabis growth, the most important factor to consider is drainage. Cannabis plants have a shallow root system and need soil that drains quickly to prevent root rot and other bacterial diseases. The ideal soil should be light, airy, and well-draining, with a pH of 6.5-7.5.

A mixture of organic soil and perlite or vermiculite will provide the best drainage. Soil drainage is an extremely important factor to consider when selecting soil for outdoor cannabis growth.

This is because it affects the water that is available for the plant and how quickly the water can move through the soil.

When it is too wet, the roots can become waterlogged and rot, while if it is too dry, the plant will not receive enough water to survive.

For cannabis to thrive, it needs soil that drains well. This means that the soil should be able to absorb and release water relatively quickly. If the soil does not drain quickly enough, the roots of the cannabis plant can become waterlogged and rot.

In addition, waterlogged soils can lead to nutrient deficiencies and other issues that will lead to poor growth and development.

The best way to test soil drainage is to fill a container with soil and then add water. If the water stays on the surface and does not soak into the soil, then the soil may be too compacted and will not drain well. If the water soaks into the soil quickly and is gone within a few minutes, then the soil is likely to have good drainage.

In addition to testing the soil drainage, it is also important to consider the type of soil that is being used. Clay soils are generally not ideal for cannabis as they are very dense and do not allow for much water and nutrient movement. On the other hand, sandy soils are ideal as they allow for more water and nutrient movement.

Loam soils (a combination of clay, sand, and organic matter) are usually the best choice for cannabis growth as they provide good drainage and nutrient retention.

Overall, soil drainage is an important factor to consider when selecting soil for outdoor cannabis growth. Poor drainage can lead to waterlogged roots, nutrient deficiencies, and other issues that will lead to poor growth and development. On the other hand, good drainage can ensure that the roots get enough water and nutrients and that the plant can develop properly.

The next factor to consider is soil texture. It is an important factor when selecting soil for outdoor cannabis growth because the texture of the soil affects the availability of nutrients, water, and air to the plant, which can all have an impact on the health of the plant, and ultimately the yield and quality of the final product.

Soil texture is determined by the size and composition of the particles that make up the soil. Generally, loamy soil with a mix of sand, silt, and clay particles is the most desirable for outdoor cannabis growth.

Loamy soil is the most ideal type of soil for outdoor cannabis growth because it can provide the plant with adequate levels of nutrients, water, and air. This is due to its mix of sand, silt, and clay particles.

The sand particles provide good drainage, allowing excess water to easily escape the soil.

The silt particles are small and provide good aeration, which helps to allow oxygen to reach the plant's roots. The clay particles are very small and hold onto nutrients, which helps to ensure that the nutrient levels in the soil remain consistent.

When selecting soil for outdoor cannabis growth, it is important to check the texture of the soil to ensure that it is not too sandy or too clay-like. If the soil is too sandy, it can lead to poor drainage and an inadequate amount of nutrients for the plant to thrive.

If the soil is too clay-like, it can lead to poor aeration and an inadequate amount of oxygen for the plant to absorb. In either case, it can lead to stunted growth and decreased yields.

The third factor to consider is soil pH. Soil pH is an important factor when selecting the soil for outdoor cannabis growth, as it affects the availability of nutrients to the plant, microbial activity in the soil, and the amount of water the soil can hold.

A soil pH of 6.0 - 7.0 is ideal for outdoor cannabis growth, as it is within the ideal range for cannabis nutrient uptake.

Soil pH also affects the microbial activity in the soil, as microbes require a certain pH range to thrive.

If the soil pH is too high, the microbes may not be able to survive, thus limiting the plant's access to beneficial microbes, such as nitrogen-fixing bacteria, which can help the plant to access nitrogen from the soil.

Additionally, soil pH can affect the amount of water the soil can hold, which is important for outdoor cannabis growth. If the soil pH is too high, the soil can become waterlogged, leading to root and plant death. Too low of a pH can lead to a soil that is too dry, which can also stunt root and plant growth.

Therefore, it is important to select soil with a pH that is within the ideal range for outdoor cannabis growth.

The next factor to consider when selecting the right soil for outdoor cannabis is nutrient content. The nutrient content is an important factor when selecting soil for outdoor cannabis growth because it determines the number of nutrients available to the plant.

The nutrient content is determined by the levels of nitrogen, phosphorus, potassium, and other essential minerals in the soil. The availability of these nutrients is essential for healthy plant growth and development.

Nitrogen is necessary for the production of chlorophyll, which is essential for photosynthesis. It is also a key component in the formation of proteins, which are essential for the growth and development of new cells.

Phosphorus is essential for the proper development of roots and flowering, as well as the production of energy. Potassium is important for the development of a strong immune system and helps to regulate the plant's water balance. Other minerals, such as calcium, zinc, and iron, are also important for the health of the plant.

When selecting the soil for outdoor cannabis growth, the soil should be tested for its nutrient content. A soil that is low in nutrients will not be able to provide the necessary nutrients for healthy plant growth. If the soil is too high in nutrients, it can lead to nutrient burn and other problems.

The ideal soil for outdoor cannabis growth should be able to provide the necessary nutrients for healthy plant growth, while not being too high in nutrients that could cause problems. Here are other factors to consider:

- Structure: The soil should be able to hold its shape and have a good crumb structure. This will allow for proper root growth and aeration.

- Organic matter: Adding organic matter such as compost or aged manure will help to improve the fertility of the soil and also provide additional nutrients.

- Aeration: The soil should be well aerated to allow for oxygen to reach the roots. This can be achieved by adding perlite or compost to the soil.

- Water retention: The soil should have good water retention to ensure that the roots don't dry out. This can be achieved by adding organic matter such as peat moss to the soil.

- Nutrient availability: The soil should have the necessary nutrients available for plant growth. This can be achieved by adding a slow-release fertilizer to the soil.
- Weed control: The soil should be free of weeds as they can compete with the cannabis plant for nutrients and water. This can be achieved by using a pre-emergent herbicide.

Finally, it is important to consider the climate you are growing in when selecting soil for outdoor cannabis growth. Soil with a high organic matter content will retain more moisture in warmer climates, while soil with a higher sand content will help plants to better tolerate cold temperatures in cooler climates.

By taking all of these factors into consideration, you can ensure that you are selecting the right soil for outdoor cannabis growth and getting the most out of your plants.

With the right soil, you can give your plants the best chance to thrive and produce a bountiful yield of buds.

CHAPTER 5

Nutrient Requirements for Outdoor Cannabis Growth

Growing cannabis outdoors requires a delicate balance of nutrients. Properly managing your plants' nutrient levels is essential to getting the best results from your outdoor crop. Here are some requirements:

1. Soil Quality: Healthy soil is essential for growing cannabis outdoors. Cannabis plants need soil that is nutrient-rich and has a pH that is slightly acidic to neutral (6.0-7.0). Plants also need access to air and water, so the soil must be well-drained and aerated.

2. Sunlight: Cannabis plants need direct access to sunlight to produce energy and grow. Marijuana plants should be exposed to eight to twelve hours of direct sunlight per day.

3. Water: Water is essential for all living things, including cannabis plants. Water helps plants absorb nutrients and regulate temperature. Cannabis plants need a consistent supply of water, so it's important to check your plants daily and water them as necessary.

4. Nutrients: Nutrients are essential for cannabis plants to grow. Nitrogen, phosphorus, and potassium are the three primary nutrients that cannabis plants need to thrive. Secondary nutrients such as calcium, magnesium, and sulfur are also important for plant health.

5. Temperature: Temperature is an important factor for outdoor cannabis growth. Cannabis plants prefer temperatures between 68-77 degrees Fahrenheit (20-25 degrees Celsius). If the temperature drops below freezing, the plants can be damaged or killed.

6. Humidity: Humidity is important for cannabis plants to absorb nutrients from the soil and to prevent powdery mildew and other types of mold. The ideal humidity level for cannabis growth is between 40-50%.

7. Pests and Disease: Pests and disease can be a major problem for outdoor cannabis growers. Regular monitoring of your plants for signs of infestation, as well as proper sanitation and prevention methods, can help to keep your plants healthy.

By following these guidelines, you can provide your outdoor cannabis crop with the nutrients it needs to produce a healthy and abundant harvest.

With proper nutrition, your plants will be able to reach their full potential and reward you with a bountiful crop.

CHAPTER 6

Watering and Feeding Your Outdoor Cannabis Plants

Watering and feeding your outdoor Cannabis plants is a crucial part of successful Cannabis cultivation. Cannabis plants require regular watering and fertilization to reach their full potential. Without proper care, your plants may fail to reach their full potential, or worse, they may die.

Watering your outdoor cannabis plants is incredibly important for their growth and development. Providing your plants with the right amount of water is critical for their health, as it helps ensure they get the nutrients and minerals they need to grow strong and healthy.

Too little water can lead to wilting and nutrient deficiencies, while too much water can cause root rot and other issues.

The key is to provide your cannabis plants with the right amount of water at the right times. This will help ensure they get the moisture they need without becoming waterlogged or overly dry.

It also helps to create an environment with the right balance of oxygen, carbon dioxide, and other elements, which can help them to thrive.

When it comes to watering your outdoor cannabis plants, you should pay attention to the weather and soil conditions. When the weather is hot and dry, your plants will need more water than usual.

In contrast, if the weather is cool and wet, you should reduce the amount of water you give your plants. Additionally, you should also monitor the soil moisture levels and adjust your watering schedule accordingly.

Finally, you should also make sure to prune your outdoor cannabis plants regularly. Pruning helps to maintain a healthy plant structure and encourages healthy growth.

When pruning your plants, you should focus on removing any dead or dying leaves and stems. This will help ensure your plants are getting the necessary air and light they need to thrive.

Feeding your outdoor Cannabis plants is incredibly important for the health, growth, and quality of their buds. Without the proper nutrients, plants won't be able to photosynthesize, meaning they won't be able to thrive and grow to their full potential.

Ensuring that your Cannabis plants get the proper nutrients is key to a successful harvest. Nitrogen, phosphorus, and potassium are the three nutrients that are most crucial for cannabis plants. These three elements are essential for the plant's growth and health.

Nitrogen helps the plant grow larger and greener, phosphorus helps with root and flower development, and potassium helps with overall plant health.

In addition to these key nutrients, Cannabis plants also need other essential minerals such as calcium, magnesium, sulfur, and iron. Calcium is important for root and cell growth, magnesium helps with chlorophyll production, sulfur is needed for protein synthesis, and iron helps with oxygen transport.

Providing your Cannabis plants with the right nutrients is essential for their health and growth. The best way to ensure that they are getting the nutrients they need is to use specially formulated Cannabis fertilizer. These fertilizers are designed to provide the exact amount of nutrients your Cannabis plants need for optimal growth.

It is important to note that Cannabis plants need lots of sunlight to thrive. Make sure that your plants have access to plenty of direct sunlight during the day.

Feeding your outdoor Cannabis plants is an essential part of their overall health. By providing the right nutrients, watering correctly, and giving them plenty of sunlight, you can ensure that your Cannabis plants will be healthy and produce quality buds.

Finally, it is important to monitor your plants for signs of stress or any potential problems. Inspect your plants regularly and watch out for any signs of disease or insect infestations.

CHAPTER 7

Understanding Light and Shade Requirements for Outdoor Cannabis Growth

Understanding light and shade requirements for outdoor cannabis growth are essential for the successful cultivation of cannabis. Cannabis is a photosensitive plant, meaning it requires a certain amount of light and shade to thrive in its environment.

Sunlight is the primary source of energy for cannabis plants, providing them with the necessary energy to complete their life cycle.

Without adequate light, the plant will not be able to produce the necessary compounds required for growth and development. The intensity of sunlight and the type of light (direct or indirect) will determine the amount of light needed for successful growth.

Producing high-quality and strong cannabis requires an understanding of the lighting needs for outdoor cannabis development. Different types of cannabis have different light requirements, so it is important to understand the specific requirements for the particular strain you are growing.

Light is one of the most important components for successful cannabis cultivation. The total amount of light a cannabis plant receives will determine its growth rate, yield, and potency. Cannabis plants require a certain amount of light to photosynthesize and produce the energy needed for growth and flowering.

The type of light that a cannabis plant needs depends on the strain and its stage of growth. Different light sources can be used to provide the required amount of light. Natural sunlight is the best source of light, but artificial lighting may also be used.

For seedlings and clones, a bright, indirect light is best. This can be achieved by placing the plants in a sunny window or by using artificial lighting. As the plants begin to grow, they will need more light to produce larger leaves and stems.

During the vegetative stage, cannabis plants will need at least 16 hours of light per day. The amount of light should be increased as the plants grow taller, as this will increase the rate of photosynthesis and stimulate growth.

When the flowering stage begins, the amount of light should be reduced to 12 hours per day, as too much light during this stage can reduce the potency of the buds. Artificial lighting can be used to supplement natural light during this stage.

The type of light used is also important. Cannabis plants prefer full-spectrum light, which mimics the light of the sun. LED lighting is becoming increasingly popular, as it is more energy efficient and produces less heat than traditional light sources.

When it comes to growing cannabis outdoors, understanding shade requirements is critical. Shade can be an important factor in controlling temperatures during the hot summer months, preventing sunburn and other damage to your plants, and helping to keep them healthy and producing good yields.

First, let's talk about the basics of shade for outdoor cannabis plants. Cannabis plants prefer to be in areas of partial shade, meaning not too much and not too little. They need just enough shade to protect them from the sun's harsh rays and keep the temperatures more moderate.

When selecting the amount of shade for your outdoor cannabis plants, it's important to consider the amount of sunlight they will receive during the day and the time of year. If your plants are in an area that receives too much sun, they may suffer from sunburn and other damage.

When it comes to the type of shade, there are a few options. You can use shade cloth to cover the plants, or you can plant trees next to the plants to provide natural shade.

Shadecloth is a great option because it is easy to install and adjust as needed.

If you decide to use trees, you need to make sure they are not too close to the plants (as they will block out the sun) or too far away (as they won't provide adequate shade).

In addition to providing shade, you also need to make sure your outdoor cannabis plants are receiving adequate air circulation. Good air circulation is key to preventing diseases and mold, and it also keeps the temperatures more moderate.

You can ensure good air circulation by planting your plants in an area with good airflow or by using fans or other air circulation devices.

Finally, it's important to remember that outdoor cannabis plants need more water than those grown indoors. Make sure you are providing enough water to keep the soil moist and consider using a drip irrigation system to ensure the plants are getting enough water.

Understanding shade requirements for outdoor cannabis growth is key to ensuring your plants are healthy and producing good yields. Make sure you are providing enough shade, air circulation, and water to keep your plants happy and thriving.

The amount of shade required will vary depending on the particular strain of cannabis and the climate in which it is grown. In general, it is best to provide enough shade to reduce direct sunlight by at least 50% during the flowering stage. In some cases, a greater amount of shade may be required if temperatures are particularly high.

CHAPTER 8

Dealing with Outdoor Cannabis Pests and Diseases

Dealing with outdoor cannabis pests can be a tricky, but necessary task for any outdoor cannabis grower. Pests can cause significant damage to plants, so proper prevention and management are key to ensuring a successful harvest.

The most common outdoor cannabis pests are aphids, mites, caterpillars, beetles, and slugs. These pests feed on the leaves, stems, and flowers of cannabis plants, which can cause significant damage to the plants and ruin a harvest.

The first step in dealing with outdoor cannabis pests is prevention. Properly caring for plants is the best way to avoid any pest infestations. Make sure to keep cannabis plants well irrigated and fertilized, as this will help the plants to stay healthy and less susceptible to pest infestations.

It's also important to keep the area around the plants clean and clear of debris and weeds, as these can act as breeding grounds for pests. It's important to regularly inspect plants for signs of pests, as early detection is key to successful pest management.

Once an infestation has occurred, there are several methods of pest management. The most effective method is using chemical pesticides, which can be applied directly to the affected plants. These should be used with caution, as they can also harm beneficial insects and pollinators. Organic pesticides, such as neem oil, can also be used, however, they may not be as effective as chemical pesticides.

Another method of pest management is introducing beneficial insects, such as ladybugs, lacewings, and wasps. These predators will help to keep pest populations in check, however, they may not be able to eliminate an infestation.

Finally, it's important to keep in mind that prevention is the key to successful pest management. If preventive measures are taken, it's much easier to avoid an infestation and ensure a successful harvest.

While dealing with cannabis disease, It might be a hard task to combat outdoor cannabis disease, but it is essential for successful harvests. This is because diseases can quickly spread and damage the quality of the crop. Fortunately, several strategies can help manage and prevent common cannabis diseases.

The first step in dealing with outdoor cannabis disease is to know what to look for. Many diseases can be identified by the symptoms they cause, such as discolored or wilting leaves, fungal spots or circles, or yellowing of the foliage.

Once the disease is identified, it is important to understand its life cycle, so that the best course of action can be taken.

The next step is to take preventative measures. This includes keeping the growing area clean and free from debris, ensuring adequate air circulation, and avoiding overcrowding. In addition, it is important to use disease-resistant varieties and to rotate crops to reduce the chance of re-infection.

Once a disease is identified, it is important to take immediate action. This may include removing the affected plant or plants, pruning away diseased or dying branches, or applying an appropriate fungicide. In cases of severe infection, it may be necessary to destroy the entire crop.

Finally, it is important to keep records of your growing practices, so that you can detect diseases early and take action quickly. This includes keeping track of temperatures, humidity, and other environmental conditions, as well as the type of soil and fertilizers used.

By following these steps, you can effectively manage and prevent outdoor cannabis pests and diseases. With careful monitoring and diligent care, you can ensure a healthy and successful harvest.

CHAPTER 9

Plant Training Techniques for Maximum Outdoor Cannabis Yields

Plant training techniques for maximum outdoor cannabis yields involve manipulating the growth of the plant to achieve the desired results. This can include manipulating the size, shape, and structure of the plant to optimize light exposure, airflow, and nutrient availability. By optimizing these factors, cannabis plants can reach their highest potential for growth and yield. Here are some techniques:

1. Topping: Topping is a plant training technique that is used to control the height of a cannabis plant, as well as to increase the number of buds produced. Topping involves cutting off the top of the plant just above the fifth node. This encourages the plant to branch out and create multiple colas (flowering sites).

2. Fimming: Fimming is a more extreme version of topping. It involves cutting off the top of the plant above the fourth node and encourages the plant to produce four main colas instead of two.

3. Low-Stress Training (LST): Low-Stress Training is a technique used to redirect the energy of the plant to produce more buds.

It involves gently bending and tying down the branches of the plant to create an even canopy. This allows light to reach all parts of the plant more evenly, resulting in better-quality buds.

4. Super cropping: Super cropping is a technique used to increase bud production on the plant. It involves gently pinching, or "crimping", the stem of the plant to break the cell walls of the stem and increase the flow of nutrients to the buds.

5. Defoliation: Defoliation is a technique used to increase airflow and light penetration throughout the plant. It involves removing some of the larger fan leaves from the plant to allow light to reach lower branches and buds.

6. Screen of Green (ScrOG): Screen of Green (ScrOG) is a technique used to further increase bud production on the plant. It involves using a trellis system, or "screen", to support the branches of the plant and to redirect energy to the lower buds. This increases the number of buds produced and makes the most of the available light.

7. Topping and Fimming with Low-Stress Training (TF-LST): Topping and Fimming with Low-Stress Training is a technique that combines the topping, fimming, and low-stress training techniques to further increase bud production on the plant.

This technique involves topping and fimming the plant and then using LST to further shape the plant and encourage even more branching.

8. High-Stress Training (HST): High-Stress Training is a technique that involves more extreme forms of plant training, such as topping multiple times, removing large amounts of foliage, and cutting into the stem of the plant. This technique can be stressful for the plant but can result in increased yields if done correctly.

9. Staking: Staking is a technique used to support the branches of the plant and prevent them from breaking due to the weight of the buds. Stakes are typically placed around the base of the plant and the branches are then tied to the stakes. This helps to evenly distribute the weight of the plant and prevents the branches from snapping.

10. Prunning: Prunning is a technique used to increase the number of buds produced on the plant. It involves running a line of string between two poles, or "P's", and tying the branches of the plant to the string. This creates an even canopy and allows light to reach all parts of the plant more evenly.

11. Trellising: Trellising is a technique used to support the plant and provide it with a solid structure. This technique involves constructing a network of strings between poles, or "T's", and tying the branches of the plant to the trellis.

This allows the plant to receive even light distribution and support for its branches and buds.

11. Sea of Green (SOG): Sea of Green (SOG) is a technique used to maximize yields in a small area. This technique involves planting multiple small plants close together to create a "sea" of cannabis plants. This allows the plants to be evenly spaced and receive even light distribution, resulting in increased yields.

The most common methods for manipulating the growth of cannabis plants for maximum outdoor yields include pruning, staking, topping, and trellising. When using these techniques, it is important to understand how each one will affect the growth of the plant and the resulting yield.

Pruning can be beneficial for increasing light exposure and airflow, however, it can also cause the plant to produce less overall biomass.

Staking can improve the structure and stability of the plant, but can also limit the height that the plant can reach.

Topping can increase yield potential, but it can also cause the plant to become vulnerable to disease or pests.

Trellising can increase the surface area of the plant that is exposed to light, however, it can also be difficult to maintain and take up a lot of space.

Therefore, when using these techniques to maximize outdoor cannabis yields, it is important to understand the pros and cons of each one and how they will affect the growth and yield of the plant.

By carefully considering each technique and taking into account the needs of the plant, growers can achieve maximum outdoor yields with the help of these plant training techniques.

CHAPTER 10

Harvesting, Drying, and Curing Outdoor Cannabis

Harvesting outdoor Cannabis is a process that requires knowledge, skill, and patience. It involves a few steps that must be carefully followed to ensure the highest quality crop.

The first step in harvesting outdoor Cannabis is to identify the plants that are ready for harvest. This is done by looking at the trichomes which are small, sticky crystals located on the flowers, leaves, and stems of the plant. When the trichomes are clear and cloudy, the plant is considered ready to be harvested.

The next step is to trim the buds and leaves. This involves removing the large fan leaves and trimming the flowers and small leaves to the desired size. This step is important to ensure good air circulation and light penetration to the buds.

Next, the buds need to be dried and cured. This is done by hanging the buds in a cool, dark, and well-ventilated area. The drying and curing process can take anywhere from one to four weeks, depending on the humidity and temperature of the room. During this time, the buds will slowly lose moisture and become more potent.

Finally, the buds need to be stored in a cool, dry place. This will ensure that the buds retain their potency and flavor for a longer period. It is also important to keep the buds away from light, as this can cause the buds to lose their potency.

Harvesting outdoor Cannabis is a long and delicate process, but when done correctly, it will produce high-quality buds that will impress and delight your customers. With the right knowledge and skill, you can ensure that your harvest is successful and your customers are satisfied.

Drying outdoor cannabis after harvesting is an important step in the cultivation process. It requires several measures to make sure the buds are cured properly and are of the best quality.

The drying process should begin as soon as the harvest is over. It is important to make sure the plants are completely dry before storing them. If moisture remains in the cannabis, it can lead to mold, mildew, and other issues that can cause the buds to become unusable.

The best way to dry outdoor cannabis is to hang it in a warm, dry, and well-ventilated area. This will help to keep the buds from molding during the drying process. Make sure to hang the plants upside down so that the stems are facing down, this will help to ensure that the buds dry evenly.

Once the buds are completely dry, it is important to keep them in an airtight container. This will help to keep the buds from drying out too quickly and from becoming overly dry. Make sure to store the cannabis in a dark and cool area.

The drying process should take about two weeks and it is important to keep an eye on the buds throughout the process. Once the buds are completely dry, it is important to trim them and remove any leaves or stems that remain. This will ensure that the buds are of the highest quality and are ready for use.

Finally, make sure to store the cannabis in a dark and cool area. This will help to preserve the buds and keep them from spoiling. Drying outdoor cannabis after harvesting is an important step in the cultivation process and it is important to make sure it is done correctly for the best results.

The next step after drying is Curing. Curing outdoor cannabis is an essential step in the cultivation process. It is the final step before the cannabis is ready to be enjoyed, and it helps to ensure that the bud is of the highest quality possible.

Curing outdoor cannabis is an art and a science. It takes patience and knowledge to get it right. The process begins with harvesting the cannabis. After harvesting, it's important to dry the buds slowly, in a dark and well-ventilated area.

This will help to preserve the terpenes, cannabinoids, and other compounds in the cannabis, and it will also help to prevent mold and mildew from forming.

Once the buds are dried, the next step is to cure them. This is done by storing the buds in airtight containers and allowing them to cure over time.

This process helps to break down the chlorophyll and other impurities in the buds, and it also helps to further preserve the terpenes and cannabinoids. It is important to ensure that the containers are kept away from light and moisture, as this will help to maintain the quality of the buds.

During the curing process, the buds should be inspected regularly. This will help to ensure that the buds are not too dry or too moist, as this can affect the taste and quality of the cannabis. It is also important to monitor the containers for mold and mildew, as this can ruin the buds.

Once the buds have been cured, they will be ready for use. At this point, they can be broken up and rolled into joints, or they can be used to make edibles, concentrates, and other products.

The curing process helps to ensure that the cannabis is of the highest quality, and it also helps to preserve the flavors and aromas of the buds.

CHAPTER 11

Common Outdoor Cannabis Growing Mistakes to Avoid

1. Planting Too Early or Too Late: Planting too early or too late can lead to stunted growth, decreased yield, and even plant death. When growing outdoors, the best time to plant is usually in late spring or early summer, when the soil has warmed up and the days are getting longer.

2. Overwatering: Overwatering is a common mistake when growing cannabis outdoors. Too much water can drown the roots and cause root rot, which can lead to stunted growth and even death of the plants. It's important to check the soil around the roots for moisture before watering and to give the plants enough time to absorb the water before adding more.

3. Not Pruning: Pruning is an essential part of outdoor cannabis growing. It helps to keep the plants healthy, increases airflow, and encourages new growth. Pruning also helps to maintain the shape of the plant and encourages the growth of bigger, healthier buds.

4. Not Protecting Plants From Pests: Pests can be a major problem when growing cannabis outdoors. Common pests such as aphids, whiteflies, and spider mites can cause

damage to the leaves and buds of the plant. To prevent this, it is important to inspect the plants regularly and take steps to protect them from pests, such as using insecticidal soap or neem oil.

5. Not Testing the Soil: Testing the soil before planting is an important step when growing cannabis outdoors. The soil should be tested for pH, nutrients, and other factors to ensure it is suitable for growing cannabis. If the soil is not suitable, the plants may suffer from nutrient deficiencies and other issues that can lead to poor growth and low yield.

6. Not Providing Adequate Lighting: Cannabis plants need a lot of light to thrive. When growing outdoors, it is important to make sure that the plants are getting enough light by positioning them in an area that gets plenty of direct sunlight each day. If the plants are not getting enough light, they may suffer from slow growth, low yield, and other issues.

6. Not Training the Plants: Training the plants is an important part of outdoor cannabis growing. Training techniques such as LST (Low-Stress Training) and SCROG (Screen of Green) can help maximize the yield of the plants and encourage healthy growth.

7. Not Setting Up Proper Security: Outdoor cannabis plants can be vulnerable to theft and vandalism. It is important to set up adequate security to protect the plants from

unwanted visitors. This may include fencing, motion sensors, and security cameras.

8. Not Taking Care of the Soil: Taking care of the soil is an important part of outdoor cannabis growing. The soil should be amended with organic matter to ensure it is healthy and full of nutrients. The soil should also be regularly monitored to check for pests, fungi, and other issues.

9. Not Utilizing Proper Harvesting Techniques: Proper harvesting techniques are essential for getting the most out of an outdoor cannabis crop. It is important to harvest the plants when the buds are ripe and to use the right techniques to ensure that the buds are dry and cured properly.

CHAPTER 12

Troubleshooting Common Outdoor Cannabis Growing Problems

Growing cannabis outdoors can be a rewarding experience, but it can also be fraught with problems. Common outdoor cannabis growing problems can range from nutrient deficiencies to pests, weather-related issues, and more. Here are some of the most common outdoor cannabis growing problems and tips for troubleshooting them:

1. Nutrient Deficiencies: One of the most common outdoor cannabis growing problems is nutrient deficiencies. Nutrient deficiencies can be caused by incorrect soil pH levels, incorrect nutrient ratios, or simply a lack of nutrients. To troubleshoot nutrient deficiencies, make sure the soil pH is between 6.0 and 6.5 for optimal cannabis growth. Additionally, make sure to use a quality fertilizer with the specific nutrient ratios that your cannabis plants need.

2. Pests: Another common outdoor cannabis growing problem is pest damage. Pest damage can range from caterpillars to spider mites and can be difficult to get rid of. To prevent and/or treat pest damage, make sure to regularly inspect your plants and remove any pests you find. You can

also use products like neem oil or insecticidal soaps to help eradicate pests.

3. Weather-Related Problems: Weather-related problems can also be one of the most common outdoor cannabis growing problems. Cannabis plants are sensitive to temperature and humidity, so it's important to make sure they are getting the right amount of sunlight and protection from extreme weather conditions. To prevent and/or treat weather-related problems, make sure to provide shade during the hottest parts of the day, and make sure your plants are getting enough water. Additionally, you can use products like row covers to protect your plants from extreme temperatures.

4. Light Burn: Light burn is another one of the most common outdoor cannabis growing problems. Light burn is caused by overexposure to direct sunlight and can cause yellowing and burning of the leaves. To prevent and/or treat light burns, make sure to provide your plants with plenty of shade during the hottest parts of the day. Additionally, make sure to keep your plants away from direct light when possible.

5. Root Rot: Root rot is a common problem for outdoor cannabis plants. Root rot is caused by over-watering and/or poor drainage, and can be difficult to treat. To prevent and/or treat root rot, make sure to water your plants correctly and use quality soil with good drainage.

Additionally, make sure to check the roots of your plants regularly to make sure they are healthy.

In conclusion, troubleshooting common outdoor cannabis growing problems can be tricky, but it's important to be aware of the potential issues and take steps to prevent them. By following the tips outlined above, you can ensure that your outdoor cannabis plants stay healthy and productive.

CHAPTER 13

Maximizing Your Outdoor Cannabis Growing Experience

Maximizing your outdoor cannabis growing experience begins with the right location. The ideal spot provides your plants with plenty of direct sunlight, as well as some protection from the elements.

When selecting a spot, look for an area with a minimum of six hours of direct sunlight per day, and that is sheltered from strong winds and other harsh weather.

Once you've found the perfect spot, the next step is to prepare your soil. Cannabis plants need nutrient-rich soil that is well-draining and airy, so be sure to amend your soil with compost or other organic matter before planting.

Next, you'll need to select the right type of cannabis strain for your location. There are many strains available, and each one will grow best in different conditions. Take into account the climate, sunlight, and soil conditions of your location when selecting a strain.

When it comes to planting, it's important to space your plants properly. Give each plant enough room to spread out its roots and receive adequate sunlight. If you're planting multiple rows, stagger them so that the plants in each row receive equal amounts of sunlight.

Once your plants are in the ground, it's time to tend to them. Be sure to water them regularly and adequately, but don't overwater them, as this can lead to root rot. If you're growing cannabis in containers, make sure to monitor the soil moisture levels and adjust your watering schedule accordingly.

It's also important to protect your plants from pests, diseases, and other issues. Keep an eye out for any signs of disease or damage, and take the necessary steps to address the issue.

Utilize natural, organic methods whenever possible, such as companion planting and using beneficial insects to naturally control pests.

Finally, once your plants are ready to harvest, it's time to enjoy your bounty! Whether you're drying and curing your buds, or extracting essential oils, be sure to take the necessary steps to maximize the quality of your product.
By following these tips and taking the time to properly care for your plants, you'll be sure to have a successful and enjoyable outdoor cannabis growing experience.

CHAPTER 14

Conclusion

Congratulations! You have just completed a journey into the world of growing marijuana for beginners. You have learned the basics of identifying different strains, preparing your growing environment, and caring for your plants. You have learned how to harvest and enjoy the end product.

As a beginner grower, you have made tremendous progress and should be proud of your accomplishments. Growing marijuana may not be easy, but with the knowledge you have gained from this book, you are well on your way to becoming an expert grower.

The journey of growing marijuana is a continuous one. As you gain more experience, you will learn how to refine your techniques and expand your growing knowledge. You will be able to experiment with different strains and methods, as well as develop your unique style.

No matter how much experience you gain, remember to always take proper precautions and remain aware of the laws in your area. Growing marijuana is a privilege and one that should be respected.

In conclusion, growing marijuana for beginners is an enjoyable and rewarding experience. You have learned the fundamentals of growing marijuana and have the tools and knowledge to begin your journey. With patience and dedication, you will be able to achieve great success in your growing endeavors. So, don't forget to enjoy the process, and don't forget to have fun!

www.ingramcontent.com/pod-product-compliance
Lightning Source LLC
Chambersburg PA
CBHW050310220526
45465CB00005B/1933